INTERNATIONAL CENTRE FOR MECHANICAL SCIENCES

COURSES AND LECTURES - No. 118

HEINZ PARKUS
TECHNICAL UNIVERSITY OF VIENNA

MAGNETO - THERMOELASTICITY

COURSE HELD AT THE DEPARTMENT
OF MECHANICS OF SOLIDS
JUNE - JULY 1972

UDINE 1972

SPRINGER-VERLAG WIEN GMBH

Originally published by Springer—Verlag Wien—New York in 1972

ISBN 978-3-211-81134-4 ISBN 978-3-7091-2938-8 (eBook)

DOI 10.1007/978-3-7091-2938-8

PREFACE

No textbook is available at the present time dealing with the phenomenological theory of the combined effects of thermal and electromagnetic fields in solids. The present short monograph represents an attempt to fill this gap. It grew out of the lectures which I gave on the subject in the summer of 1972 at the Centre International des Sciences Mécaniques in Udine.

My sincere thanks are due to the Secretary General of CISM, Prof. L. Sobrero, and to the Rector, Prof. W. Olszak, for kindly inviting me to present these lectures and to write this little book.

Udine, June - July 1972

Introduction

The interaction between electric and magnetic fields on the one hand, and hot gases (plasma) on the other is of great practical importance. Known as "magnetogasdynamics" it has developed into a wide field of research. In contradistinction, the corresponding problem of the interaction between electromagnetic fields and solid bodies has long remained dormant. Of course, certain special effects as, for instance, piezoelectricity or photoelasticity, have been well studied and put to technological use. But it is only for a relatively recent period of time that a general theory has been developed.

The first papers on magnetoelastic interaction are due to Becker [1]. Knopoff [2] started investigating the influence of magnetic fields on the propagation of elastic waves. Following first attempts towards a theory of photoelasticity by F.E.Neumann (1841) and a theory of piezoelectricity by W.Voigt (1890), a systematic theory of the elastic dielectric with finite deformations has been worked out by Toupin [3] for the static case. Later [4], he generalized his theory to include dynamic effects. The corresponding theory of the elastic ferromagnetic body was developed by Brown Jr. [5] and, independently, by Tiersten [6]. Recently, the latter also gave a theory of the elastic dielectric [7]. Important contributions are due to Eringen who, in

1963, published a basic paper on the elastic dielectric [8]. This was followed by a series of publications on both dielectric and ferromagnetic bodies by him and his coworkers, cf. [9] and other papers.

In the following, an attempt is made to give a survey of the present state of the phenomenological theory for "slowly" moving thermoelastic bodies, i.e., with relativistic effects neglected. There are only relatively few applications of magnetoelasticity available in the literature, which include thermal effects. Most of them are concerned with wave propagation and are discussed in Chapter 4. For other applications the reader is referred to a forthcoming review paper by the present author in the ZAMM.

The list of references is supposed to serve as a representative cross-section through the corresponding literature. Completeness was neither achieved nor intended.

NOTATION

All equations are written in the international MSK system (Giorgi system). Basic units are meter m, second s, kilogram kg and Ampère A. (Volt $V = m^2 kg/As^3$, Newton $N = kg\,m/s^2$).

A	surface area m^2
$\underset{\sim}{B}$	magnetic induction, Vs/m^2
$\underset{\sim}{D}$	electric displacement, As/m^2
$\underset{\sim}{E}$	electric field strength, V/m
E	modulus of elasticity, N/m^2
F	free energy per unit of mass, Nm/kg
$\underset{\sim}{F}$	force, N
G	shear modulus, N/m^2
$\underset{\sim}{H}$	magnetic field strength, A/m
$\underset{\sim}{J}$	surface current density, A/m
$\underset{\sim}{L}$	moment, Nm
$\underset{\sim}{M}$	magnetization per unit of volume, A/m
$\mathcal{M}$	magnetization per unit of mass , $\mathcal{M} = M/\rho$
$\underset{\sim}{P}$	polarization per unit of volume, As/m^2
$\underset{\sim}{\mathcal{P}}$	polarization per unit of mass, $\mathcal{P} = P/\rho$
Q	heat flux, N/ms
S	entropy per unit of mass, $N/ms\,°K$
T	absolute temperature, $°K$, T_0 reference temperature
U	internal energy per unit of mass Nm/kg

U_e	electromagnetic energy per unit of volume, N/m^2
V	volume, m^3
V	wave speed, m/s
X_A	spatial coordinate, m,
a_{ij}	exchange tensor, N/A
c	speed of light in vacuo, m/s
$\underset{\sim}{f}$	volume force density, N/m^3
$\underset{\sim}{f}_L$	Lorentz force density, N/m^3
$\underset{\sim}{j}$	electric current density, A/m^2
k	thermal conductivity, $N/s\,^\circ K$
m	mass, kg
$\underset{\sim}{n}$	unit normal vector
r	strength of heat source distribution, $Nm/kg\,s$
t	time, s
t_{ij}	stress tensor, N/m^2
u_i	displacement vector, m
v_i	particle velocity, m/s
x_i	material coordinate, m
$x_{i,A}$	deformation gradient
α_T	coefficient of thermal expansion, $1/^\circ K$
γ	wave number, $1/m$
ε	dielectric constant, As/Vm
ε_0	dielectric constant of free space, $\varepsilon_0 = 8.859 \times 10^{-12}$
ε_{ij}	strain tensor

ϑ	temperature, °K, $\vartheta = T - T_0$
$\varkappa$	coefficient in Ohm's law, V/m °K
μ	permeability, Vs/Am
μ_0	permeability of free space, $\mu_0 = 1.257 \times 10^{-6}$
	$\varepsilon_0 \mu_0 = 1/c^2$
ν	Poisson's ratio
$\underset{\sim}{\Pi}$	measure of polarization
ρ	mass density, kg/m^3
ρ_e	charge density, As/m^3
σ	electrical conductivity A/Vm
σ_{ij}	stress tensor, N/m^2
τ_{ij}	Cauchy stress tensor, N/m^2
ω	circular frequency, 1/s
ω_{ij}	rotation tensor

Chapter 1

SOME BASIC RELATIONS

1.1 The Electromagnetic Field

The theory of electromagnetic fields is governed by the Maxwell equations (*)

(1.1.1)
$$\begin{cases} \nabla \times \mathbf{H} = \mathbf{j} + \dfrac{\partial \mathbf{D}}{\partial t}, & \nabla \times \mathbf{E} = -\dfrac{\partial \mathbf{B}}{\partial t} \\ \nabla \cdot \mathbf{D} = \rho_o, & \nabla \cdot \mathbf{B} = 0 . \end{cases}$$

The first two equations relate the electric field $\mathbf{E}$, the magnetic field $\mathbf{H}$, the electric displacement $\mathbf{D}$, the magnetic induction (or flux density) $\mathbf{B}$ and the electric current density $\mathbf{j}$. The third and fourth equation are, to some extent, consequences of the first two. ρ_o denotes the electric charge density. The fourth equation shows that no magnetic charge, i.e., no single magnetic pole exists. The del operator ∇ is defined by Eq. (1.2.4).

From the first and third equation the law of electric continuity

(1.1.2)
$$\nabla \cdot \mathbf{j} + \frac{\partial \rho_e}{\partial t} = 0$$

follows.

(*) see [10], for instance.

The integral form of Eqs. (1.1.1) may be obtained by integrating over a surface A or a volume V and using Stokes' and Gauss' theorem, respectively, as

$$\left.\begin{aligned} \oint_C \underset{\sim}{H} \cdot d\underset{\sim}{s} &= \int_A \underset{\sim}{j} \cdot d\underset{\sim}{A} + \frac{d}{dt}\int_A \underset{\sim}{D} \cdot d\underset{\sim}{A} \\ \oint_C \underset{\sim}{E} \cdot d\underset{\sim}{s} &= -\frac{d}{dt}\int_A \underset{\sim}{B} \cdot d\underset{\sim}{A} \end{aligned}\right\} \quad (1.1.3)$$

where A is a stationary surface with closed boundary curve C and directed line element $d\underset{\sim}{s}$, while $d\underset{\sim}{A} = \underset{\sim}{n}\,dA$ and $\underset{\sim}{n}$ is the unit normal of A.

Similarly,

$$\oint_{\partial V} \underset{\sim}{D} \cdot \underset{\sim}{n}\, d\partial V = \int_V \rho_e\, dV, \quad \oint_{\partial V} \underset{\sim}{B} \cdot \underset{\sim}{n}\, d\partial V = 0 \quad (1.1.4)$$

where V is a stationary volume with closed boundary surface ∂V.

Maxwell's equations are valid both inside and outside of matter. They must be supplemented by jump conditions across a surface of discontinuity, e.g., the surface of a body,

$$\left.\begin{aligned} \underset{\sim}{n} \times [\underset{\sim}{H}] = \underset{\sim}{J} - V[\underset{\sim}{D}], \quad \underset{\sim}{n} \times [\underset{\sim}{E}] = V[\underset{\sim}{B}] \\ \underset{\sim}{n} \cdot [\underset{\sim}{D}] = 0, \quad \underset{\sim}{n} \cdot [\underset{\sim}{B}] = 0 \,. \end{aligned}\right\} \quad (1.1.5)$$

Here, $[\underset{\sim}{H}] = \underset{\sim}{H}^+ - \underset{\sim}{H}^-$ denotes the jump of the vector $\underset{\sim}{H}$ across the surface of discontinuity in the direction of the normal vector $\underset{\sim}{n}$, and V denotes the component in the direction $\underset{\sim}{n}$ of the surface ve locity. The vector $\underset{\sim}{J}$ represents the density of the surface current.

In vacuum the following simple linear relations, known as Lorentz ether relations, exist between the electric field $\underset{\sim}{E}$ and the electric displacement vector $\underset{\sim}{D}$ on the one hand, and the magnetic field $\underset{\sim}{H}$ and the magnetic flux density $\underset{\sim}{B}$ on the other:

(1.1.6)
$$\underset{\sim}{D} = \varepsilon_0 \underset{\sim}{E}, \qquad \underset{\sim}{B} = \mu_0 \underset{\sim}{H}$$

where ε_0 and μ_0 are universal constants.

In matter, additional fields appear as a consequence of the polarization of the body. Eqs. (1.1.6) are then to be replaced by

(1.1.7)
$$\underset{\sim}{D} = \varepsilon_0 \underset{\sim}{E} + \underset{\sim}{P}, \qquad \underset{\sim}{B} = \mu_0 (\underset{\sim}{H} + \underset{\sim}{M})$$

where $\underset{\sim}{P}$ and $\underset{\sim}{M}$ represent electric polarization and magnetization, respectively, per unit of volume of the body.

If Eqs. (1.1.7) are substituted into Eqs. (1.1.1), one obtains

(1.1.8)
$$\begin{cases} \nabla \times \dfrac{\underset{\sim}{B}}{\mu_0} = \dot{\underset{\sim}{j}} + \varepsilon_0 \dfrac{\partial \underset{\sim}{E}}{\partial t} + \nabla \times \underset{\sim}{M} + \dfrac{\partial \underset{\sim}{P}}{\partial t}, \quad \nabla \times \underset{\sim}{E} = -\dfrac{\partial \underset{\sim}{B}}{\partial t} \\ \varepsilon_0 \nabla \cdot \underset{\sim}{E} = \rho_e - \nabla \cdot \underset{\sim}{P}, \qquad \nabla \cdot \underset{\sim}{B} = 0 . \end{cases}$$

By comparing these relations with those of free space, where $\underset{\sim}{M} = \underset{\sim}{P} = 0$, one notes that the presence of polarized matter may be interpreted as producing a polarization current $\partial \underset{\sim}{P}/\partial t$, a magnetization current $\nabla \times \underset{\sim}{M}$ and an electric charge density $- \nabla \cdot \underset{\sim}{P}$.

1.2 Moving Bodies

In the preceding section the body has been assumed to be at rest relative to the free space which we identify with an inertial frame. Consider now a particle (material point of a body) moving with velocity $\underset{\sim}{v}(\underset{\sim}{x}, t) = \dot{\underset{\sim}{x}}$, where

$$x_i = x_i(\underset{\sim}{X}, t) \tag{1.2.1}$$

is the instantaneous or spatial coordinate of the particle (Eulerian coordinate), and X_A is its initial or material coordinate (Lagrangian coordinate). A dot denotes time derivative,

$$\dot{f}_i \equiv \frac{d f_i}{dt} . \tag{1.2.2}$$

Now, while the rate of change of a vector $\underset{\sim}{f}$ relative to a fixed particle is given by $\partial \underset{\sim}{f}/dt$, the rate of change relative to the moving particle is given by the convected time flux, defined as (*)

$$\frac{d_c \underset{\sim}{f}}{dt} = \frac{\partial \underset{\sim}{f}}{\partial t} + \underset{\sim}{v}(\nabla \cdot \underset{\sim}{f}) + \nabla \times (\underset{\sim}{f} \times \underset{\sim}{v}) \tag{1.2.3a}$$

(*) see [11], p. 448 and 675.

or, equivalently,

(1.2.3b)
$$\frac{d_c \underset{\sim}{f}}{dt} = \frac{\partial \underset{\sim}{f}}{\partial t} + (\underset{\sim}{v} \cdot \nabla)\underset{\sim}{f} + \underset{\sim}{f}(\nabla \cdot \underset{\sim}{v}) - (\underset{\sim}{f} \cdot \nabla)\underset{\sim}{v}$$

where

(1.2.4)
$$\nabla_i \equiv \partial/\partial x_i \equiv (\ )_{,i} \,.$$

In components, Eq. (1.2.3) reads

(1.2.5)
$$\frac{d_c f_i}{dt} = \frac{\partial f_i}{\partial t} + v_j f_{i,j} + f_i v_{j,j} - f_j v_{i,j} \,.$$

For a scalar a one has instead

(1.2.6)
$$\frac{d_c a}{dt} = \frac{\partial a}{dt} + (\underset{\sim}{v} \cdot \nabla) a + a(\nabla \cdot \underset{\sim}{v}) \,.$$

The values of $\underset{\sim}{E}, \underset{\sim}{D}, \underset{\sim}{B}, \underset{\sim}{H}, \underset{\sim}{j}$ and ρ_e in Eqs. (1.1.1) are those as observed from a frame fixed in space ("laboratory frame"). Let E', D', B', H', j' and ρ_e' be the corresponding quantities as observed from the moving body. These quantities too must obey Maxwell's equations, i.e., we must have

(1.2.7)
$$\begin{cases} \nabla \times \underset{\sim}{H}' = \underset{\sim}{j}' + \dfrac{d_c \underset{\sim}{D}'}{dt} \,, & \nabla \times \underset{\sim}{E}' = - \dfrac{d_c \underset{\sim}{B}'}{dt} \\[2ex] \nabla \cdot \underset{\sim}{D}' = \rho_e' \,, & \nabla \cdot \underset{\sim}{B}' = 0 \,. \end{cases}$$

Substituting from Eq. (1.2.3a) into the first of Eqs. (1.2.7) and comparing with Eq. (1.1.1), we get, using $\nabla \cdot \underset{\sim}{D}' = \rho_e'$,

$$\nabla \times (\underset{\sim}{H}' + \underset{\sim}{v} \times \underset{\sim}{D}') - \underset{\sim}{j}' - \rho_e' \underset{\sim}{v} - \frac{\partial \underset{\sim}{D}'}{\partial t} = \nabla \times \underset{\sim}{H} - \underset{\sim}{j} - \frac{\partial \underset{\sim}{D}}{\partial t}$$

and hence

$$\underset{\sim}{H}' + \underset{\sim}{v} \times \underset{\sim}{D}' = \underset{\sim}{H}, \qquad \underset{\sim}{D}' = \underset{\sim}{D}, \quad \underset{\sim}{j}' + \rho_e' \underset{\sim}{v} = \underset{\sim}{j}.$$

Similarly, from the second of Eqs. (1.2.7),

$$\underset{\sim}{E}' - \underset{\sim}{v} \times \underset{\sim}{B}' = \underset{\sim}{E}, \quad \underset{\sim}{B}' = \underset{\sim}{B}, \qquad \rho_e' = \rho_e .$$

Summing up we have, therefore, the following relations,

$$\left.\begin{array}{l} \underset{\sim}{E}' = \underset{\sim}{E} + \underset{\sim}{v} \times \underset{\sim}{B} \quad , \quad \underset{\sim}{H}' = \underset{\sim}{H} - \underset{\sim}{v} \times \underset{\sim}{D} \\ \underset{\sim}{j}' = \underset{\sim}{j} - \rho_e \underset{\sim}{v} , \end{array}\right\} . \tag{1.2.8}$$

All other quantities remain unchanged under the motion.

The last of Eqs. (1.2.8) is intuitively obvious: since a moving charge represents a current, the observer at rest observes a current $\underset{\sim}{j}$ which differs by $\rho_e \underset{\sim}{v}$ from the current $\underset{\sim}{j}'$ relative to the moving particle.

We note that the Lorentz ether relations (1.1.6) are not invariant under the transformation (1.2.8). The same holds true for Eqs. (1.1.7). In order to formulate them for the moving body we return to Eqs. (1.1.8). The polarization $\underset{\sim}{P}$ is moving with the body. The polarization current $\partial \underset{\sim}{P}/\partial t$ has, therefore, to be replaced by $d_c\underset{\sim}{P}/dt$. Furthermore, a term $-\underset{\sim}{v}(\nabla\cdot\underset{\sim}{P})$ has to be added, which corresponds to the moving polarization charge $-\nabla\cdot\underset{\sim}{P}$. Thus we have, instead of the first of Eqs. (1.1.8),

(1.2.9a) $$\nabla\times\frac{\underset{\sim}{B}}{\mu_0} = \dot{\underset{\sim}{j}} + \varepsilon_0\frac{\partial\underset{\sim}{E}}{\partial t} + \nabla\times\underset{\sim}{M} + \frac{d_c\underset{\sim}{P}}{dt} - \underset{\sim}{v}(\nabla\cdot\underset{\sim}{P})$$

or, using Eq. (1.2.3a),

(1.2.9b) $$\nabla\times\frac{\underset{\sim}{B}}{\mu_0} = \dot{\underset{\sim}{j}} + \varepsilon_0\frac{\partial\underset{\sim}{E}}{\partial t} + \nabla\times\underset{\sim}{M} + \frac{\partial\underset{\sim}{P}}{\partial t} + \nabla\times(\underset{\sim}{P}\times\underset{\sim}{v}).$$

The other Eqs. (1.1.8) remain unchanged.

From a comparison of Eq. (1.2.9b) and Eqs. (1.1.8) we conclude now that Eqs. (1.1.7) have to be replaced for a moving body by

(1.2.10) $$\underset{\sim}{D} = \varepsilon_0\underset{\sim}{E} + \underset{\sim}{P}, \qquad \underset{\sim}{B} = \mu_0(\underset{\sim}{H} + \underset{\sim}{M} - \underset{\sim}{v}\times\underset{\sim}{P}).$$

For an electrically and magnetically <u>linear and isotropic</u> material we have the constitutive relations

(1.2.11) $$\underset{\sim}{P} = \chi_e\underset{\sim}{E}, \qquad \underset{\sim}{M} = \chi_m\underset{\sim}{H}$$

where χ_e and χ_m are constants. Eqs. (1.1.7) then simplify to

(1.2.12) $$\begin{cases} \underset{\sim}{D} = \varepsilon\underset{\sim}{E}, & \underset{\sim}{B} = \mu\underset{\sim}{H} \\ \varepsilon = \varepsilon_0 + \chi_e, & \mu = \mu_0(1+\chi_m). \end{cases}$$

The constants ε and μ are called dielectric constant and permeability, respectively (*).

For the moving body, since polarization and magnetization are material quantities, carried with the body, Eqs. (1.2.11) have to be replaced by

$$\underset{\sim}{P} = \chi_e(\underset{\sim}{E} + \underset{\sim}{v} \times \underset{\sim}{B}), \quad \underset{\sim}{M} = \chi_m(\underset{\sim}{H} - \underset{\sim}{v} \times \underset{\sim}{D})$$

and Eqs. (1.2.12) transform into

$$\underset{\sim}{D} = \varepsilon\underset{\sim}{E} + \chi_e \underset{\sim}{v} \times \underset{\sim}{B}, \qquad \underset{\sim}{B} = \mu\underset{\sim}{H} - \left(\varepsilon\mu - \frac{1}{c^2}\right)\underset{\sim}{v} \times \underset{\sim}{E}. \quad (1.2.13)$$

In arriving at the second equation, quadratic and higher terms in v have been neglected, i.e.,

$$M = \chi_m(\underset{\sim}{H} - \varepsilon\underset{\sim}{v} \times \underset{\sim}{E})$$

has been written, and the relation

$$\mu_0(\varepsilon\chi_m + \chi_e) = \mu_0(\varepsilon_0\chi_m + \chi_e\chi_m + \chi_e) = \varepsilon\mu - \varepsilon_0\mu_0 = \varepsilon\mu - \frac{1}{c^2}$$

has been used.

For a derivation of these equation as the "slow motion" approximation to the exact Minkowski equations of relativistic electrodynamics of a linear dielectric ($\chi_m = 0$), see

---

(*) Still other names are in use and may be found in the literature.

[11] sect. 308, and [13], § 57.

1.3 The Poynting Vector

If the first of Maxwell's Eqs. (1.1.1) is multiplied by $\underset{\sim}{E}$, and the second by $\underset{\sim}{H}$ and subtracted, we obtain

(1.3.1)
$$\nabla \cdot \left(\underset{\sim}{E} \times \underset{\sim}{H}\right) + \underset{\sim}{j} \cdot \underset{\sim}{E} = -\left(\underset{\sim}{E} \cdot \frac{\partial \underset{\sim}{D}}{\partial t} + \underset{\sim}{H} \frac{\partial \underset{\sim}{B}}{\partial t}\right)$$

or, after integration over a volume fixed in a space, and application of Gauss' theorem,

(1.3.2)
$$\oint_{\partial V} \left(\underset{\sim}{E} \times \underset{\sim}{H}\right) \cdot \underset{\sim}{n} \, d\partial V + \int_V \underset{\sim}{j} \cdot \underset{\sim}{E} \, dV = - \int_V \left(\underset{\sim}{E} \cdot \frac{\partial \underset{\sim}{D}}{\partial t} + \underset{\sim}{H} \frac{\partial \underset{\sim}{B}}{\partial t}\right) dV .$$

All terms in this equation have the dimension of rate of energy. The quantity $\underset{\sim}{E} \times \underset{\sim}{H}$ is known as the Poynting vector. The scalar $(\underset{\sim}{E} \times \underset{\sim}{H}) \cdot \underset{\sim}{n}$ represents the flux of electromagnetic energy through the surface of the volume into the surrounding space.

Chapter 2

THERMOMECHANICS OF FERROMAGNETIC BODIES

Even in weak magnetic fields, ferromagnetic media are strongly magnetized. They are characterized by a magnetization vector $\underset{\sim}{M}$. In general, they also conduct heat and electric current, but they do not exhibit electrical polarization, i.e., $\underset{\sim}{P} = 0$.

The most important ferromagnetic materials are iron, nickel and cobalt. Nonmetallic ferrites show a similar behavior, but posses a very high electrical resistance, by a factor of several millions larger than that of metallic ferromagnetics.

2.1 The First and Second Law

The first law of thermodynamics, i.e., the energy balance, for a moving and deforming ferromagnetic body of instantaneous volume V may be written in the form, [14],

$$\frac{d}{dt}\int_V \left[\rho\left(\frac{v^2}{2} + U\right) + U_e\right] dV = \int_V (\rho r + f_i v_i)\, dV + \tag{2.1.1}$$

$$+ \oint_{\partial V} \left[\overset{*}{\tau}_{ij} v_j + a_{ij}\rho \frac{d\mathcal{M}_j}{dt} - Q_i - (\underset{\sim}{E} \times \underset{\sim}{H})_i + U_e v_i\right] n_i\, d\partial V .$$

The left-hand side represents the time rate of the total energy (kinetic, internal and electromagnetic) enclosed in V. The terms on the right-hand side are: heat production by the heat source distribution, rate of work of volume forces f_i , of surface forces $\tau^*_{ij} n_i$ and of "exchange forces" $a_{ij} n_i$, transport of heat $-Q_i n_i$ and, from Eq. (1.3.2), of electromagnetic energy $-(\underset{\sim}{E} \times \underset{\sim}{H})_i n_i$ through the surface into the body (n_i positive outwards) and, finally, the influx of electromagnetic energy $U_e v_n$ due to the motion of the body through the external electromagnetic field.

The magnetization vector density $\mathcal{M}_i$ is introduced here with reference to the unit of mass

(2.1.2) $$\rho \mathcal{M}_i = M_i ,$$

The, as yet unknown, stress tensor τ^*_{ij} contains the mechanical stress tensor τ_{ij} plus additional magnetic effects. The exchange tensor a_{ij} covers the exchange forces between the mechanical continuum and the electronic spin continuum. It, too, is unknown.

As has been pointed out already in Sect. 1.2, the motion of a particle will be described by its spatial coordinates

(2.1.3) $$x_i = x_i(\underset{\sim}{X} , t) \qquad (i = 1,2,3)$$

where X_A , $A = 1,2,3$ represents the material coordinates which initially coincide with x_i ,

(2.1.4) $$x_i(\underset{\sim}{X} , 0) = X_i .$$

The deformation gradient $x_{i,A}$ is chosen to serve as a strain measure. The particle velocity v_i is given by

$$v_i = dx_i/dt = \dot{x}_i \tag{2.1.5}$$

or, alternatively, by

$$v_i = \dot{u}_i \tag{2.1.6}$$

where $\underset{\sim}{u} = \underset{\sim}{x} - \underset{\sim}{X}$ is the displacement vector.

The second law of thermodynamics is assumed in the form of the Clausius-Duhem inequality as, [12], p. 364,

$$\frac{d}{dt}\int_m S\, dm \geqslant \int_m \frac{r}{T} dm - \oint_{\partial V} \frac{Q_i n_i}{T} d\partial V \tag{2.1.7}$$

where S denotes entropy per unit mass and T is absolute temperature.

Applying now Gauss' theorem to Eq. (2.1.1) and bearing in mind that

$$\frac{d}{dt}\int_V U_e\, dV - \oint_{\partial V} U_e v_n\, d\partial V = \int_V \frac{\partial U_e}{\partial t} dV \tag{2.1.8}$$

one obtains the differential equation form of the first law as

$$\rho \frac{d}{dt}\left(\frac{v^2}{2} + U\right) + \frac{\partial U_e}{\partial t} = \rho r + f_i v_i + \frac{\partial}{\partial x_j}\left[\overset{*}{t}_{ji} v_i + a_{ji}\rho \frac{d\mathcal{M}_i}{dt} - Q_j - (\underset{\sim}{E} \times \underset{\sim}{H})_j\right]. \tag{2.1.9}$$

Similarly for the second law, from Eq. (2.1.7),

(2.1.10) $$\rho T \frac{dS}{dt} \geqslant \rho r - Q_{i,i} + \frac{Q_i}{T} T_{,i}.$$

The free energy F per unit of mass, defined by

(2.1.11) $$F = U - TS$$

of the elastic, ferromagnetic body is assumed as a function of strain, magnetization vector and its gradient, and of temperature:

(2.1.12) $$F = F_1 \left(x_{i,A}, \mathcal{M}_i, \mathcal{M}_{i,j}, T \right).$$

We have then

(2.1.13) $$\rho \frac{dF_1}{dt} = \rho \left(\frac{\partial F_1}{\partial x_{i,A}} \frac{dx_{i,A}}{dt} + \frac{\partial F_1}{\partial \mathcal{M}_i} \frac{d\mathcal{M}_i}{dt} + \frac{\partial F_1}{\partial \mathcal{M}_{i,j}} \frac{d\mathcal{M}_{i,j}}{dt} + \frac{\partial F_1}{\partial T} \frac{dT}{dt} \right)$$

and

$$\rho \frac{\partial F_1}{\partial x_{i,A}} \frac{dx_{i,A}}{dt} = \rho \frac{\partial F_1}{\partial x_{i,A}} v_{i,A} = \rho \frac{\partial F_1}{\partial x_{i,A}} v_{i,j}\, x_{j,A} =$$

(2.1.14) $$= \frac{\partial}{\partial x_j} \left(\rho \frac{\partial F_1}{\partial x_{i,A}} x_{j,A} v_i \right) - \frac{\partial}{\partial x_j} \left(\rho \frac{\partial F_1}{\partial x_{i,A}} x_{j,A} \right) v_i$$

$$\rho \frac{\partial F_1}{\partial M_{i,j}} \frac{d\mathcal{M}_{i,j}}{dt} = \frac{\partial}{\partial x_j} \left(\rho \frac{\partial F_1}{\partial \mathcal{M}_{i,j}} \frac{d\mathcal{M}_i}{dt} \right) - \frac{\partial}{\partial x_j} \left(\rho \frac{\partial F_1}{\partial \mathcal{M}_{i,j}} \right) \frac{d\mathcal{M}_i}{dt}.$$

(2.1.15)

For the electromagnetic energy we introduce the expression

$$U_e = \frac{1}{2}\left(\varepsilon_0 E^2 + \mu_0 H^2\right) \qquad (2.1.16)$$

where $\underset{\sim}{E}$ and $\underset{\sim}{H}$ are electric and magnetic field intensity, respectively, and ε_0 and μ_0 are dielectric constant and permeability in vacuum, respectively. Then, making use of Maxwell's equations (1.1.1),

$$\nabla \times \underset{\sim}{H} = \underset{\sim}{j} + \frac{\partial \underset{\sim}{D}}{\partial t}, \qquad \nabla \times \underset{\sim}{E} = -\frac{\partial \underset{\sim}{B}}{\partial t} \qquad (2.1.17)$$

and of the constitutive relations (1.1.7) for a moving, non-polarized body

$$\underset{\sim}{D} = \varepsilon_0 \underset{\sim}{E}, \qquad \underset{\sim}{B} = \mu_0\left(\underset{\sim}{H} + \underset{\sim}{M}\right) \qquad (2.1.18)$$

we find

$$\frac{\partial U_e}{\partial t} = \varepsilon_0 E_i \frac{\partial E_i}{\partial t} + \mu_0 H_i \frac{\partial H_i}{\partial t} = -\nabla\cdot\left(\underset{\sim}{E} \times \underset{\sim}{H}\right) - j_i E_i - \mu_0 H_i \frac{\partial M_i}{\partial t}$$

$$= -\nabla\cdot\left(\underset{\sim}{E} \times \underset{\sim}{H}\right) - j_i E_i - \mu_0 H_i \rho \frac{d\mathcal{M}_i}{dt} + \mu_0\left(\rho H_i \mathcal{M}_i v_k\right)_{,k} - \mu_0 \rho \mathcal{M}_i H_{i,k} v_k .$$

(2.1.19)

In writing Eq. (2.1.19), the continuity equation

$$\frac{\partial \rho}{\partial t} + \frac{\partial}{\partial x_i}\left(\rho v_i\right) = 0 \qquad (2.1.20)$$

as well as the relation

(2.1.21)
$$\frac{d\mathcal{M}_i}{dt} = \frac{\partial \mathcal{M}_i}{\partial t} + v_j\, \mathcal{M}_{i,j}$$

have been utilized.

Another constitutive equation, Ohm's law, will also be needed (*). For an electrically isotropic body this law reads

(2.1.22)
$$\underset{\sim}{j} = \sigma(\underset{\sim}{E} + \underset{\sim}{v} \times \underset{\sim}{B} - \varkappa\, \nabla T)$$

where σ represents the conductivity. For an anisotropic medium, σ generalizes to a symmetric tensor $\underset{\sim}{S}$.

Multiplication of both sides of Eq. (2.1.22) by $\underset{\sim}{j}/\sigma$ yields

(2.1.23)
$$\frac{j^2}{\sigma} = \underset{\sim}{j} \cdot \underset{\sim}{E} - (\underset{\sim}{j} \times \underset{\sim}{B}) \cdot \underset{\sim}{v} - \varkappa\, \underset{\sim}{j} \cdot \nabla T .$$

Substitution of Eqs. (2.1.19) and (2.1.23), together with Eqs. (2.1.11), (2.1.13), (2.1.14) and (2.1.15) into equation (2.1.9) and inequality (2.1.10) renders

$$\left\{\rho \frac{d v_i}{dt} - f_i - \frac{\partial}{\partial x_j}\left(\rho \frac{\partial F_1}{\partial x_{i,A}}\, x_{j,A}\right) - (\underset{\sim}{j} \times \underset{\sim}{B})_i - \mu_0 \rho \mathcal{M}_k H_{k,i}\right\} v_i +$$

(*) See [13] , § 25.

$$+\rho\left\{\frac{\partial F_1}{\partial \mathcal{M}_i}-\frac{1}{\rho}\frac{\partial}{\partial x_j}\left(\rho\frac{\partial F_1}{\partial \mathcal{M}_{i,j}}\right)-\mu_0 H_i\right\}\frac{d\mathcal{M}_i}{dt}+\rho\left\{\frac{\partial F_1}{\partial T}+S\right\}\frac{dT}{dt}$$

$$+\frac{\partial}{\partial x_j}\left\{\left(\rho\frac{\partial F_1}{\partial x_{i,\lambda}}x_{j,\lambda}-\overset{*}{\tau}_{ji}+\mu_0\rho\mathcal{M}_s H_s\delta_{ij}\right)v_i+\rho\left(\frac{\partial F_1}{\partial \mathcal{M}_{i,j}}-\right.\right.$$

$$\left.\left.-a_{ji}\right)\frac{d\mathcal{M}_i}{dt}+Q_j\right\}+\rho T\frac{dS}{dt}-\frac{\dot{j}^2}{\sigma}-\varkappa\,\underset{\sim}{j}\cdot\nabla T-\rho r=0 \tag{2.1.24}$$

and

$$\left\{\dots\right\}v_i+\rho\left\{\dots\right\}\frac{d\mathcal{M}_i}{dt}+\rho\left\{\dots\right\}\frac{dT}{dt}+\frac{\partial}{\partial x_j}\left\{\dots\right\}$$

$$-\frac{\dot{j}^2}{\sigma}-\varkappa\,\underset{\sim}{j}\cdot\nabla T+\frac{Q_i}{T}T_{,i}\leq 0 \tag{2.1.25}$$

where the expression in braces in inequality (2.1.25) are identical with the corresponding terms in Eq. (2.1.24).

2.2 The Basic Equations

A number of conclusions may now be drawn from the first and second law in the form of relations (2.1.24) and (2.1.25). First, we note that, if we assume S independent of the temperature rate dT/dt, the coefficient of dT/dt must vanish. This yields the well-known thermodynamic relation

(2.2.1)
$$S = -\frac{\partial F_1}{\partial T}.$$

We now make use of invariance conditions under superposed rigid body motions, [15] , by first replacing v_i by $v_i + c_i$ (rigid translation) and then $v_{i,j}$ by $v_{i,j} + \Omega_{ij}$ (rigid rotation). It follows that those terms which have v_i as a factor must vanish. (*) This renders the two equations

(2.2.2)
$$\rho \dot{v}_i = f_i + \tau_{ji,j} + (\underset{\sim}{j} \times \underset{\sim}{B})_i + \mu_0 M_k H_{k,i}$$

and

(2.2.3)
$$\tau^*_{ij} = \tau_{ij} + \mu_0 M_j H_j \delta_{ij}$$

where τ_{ik}, defined by

(2.2.4)
$$\tau_{ik} = \rho x_{i,A} \frac{\partial F_1}{\partial x_{k,A}}$$

represents the Cauchy stress tensor. Eq. (2.2.2) represents the <u>equation of motion</u> while Eq. (2.2.3) determinates τ^*_{ij} in the energy equation (2.1.1).

Next we consider the terms containing $d\mathcal{M}_i/dt$ as a factor. They must vanish. Now, the magnetic equation of angular momentum (**) for the magnetic moment $\underset{\sim}{\mathcal{M}}$ per unit mass

(*) According to Eq. (2.1.22) j_i too varies with v_i . This, however, leads to a term quadratic in the velocity which is to be neglected in our "slow motion" theory.

(**) See [5] , p. 85.

reads

$$\frac{d\underset{\sim}{\mathcal{M}}}{dt} = \gamma \underset{\sim}{\mathcal{M}} \times \underset{\sim}{H}_{eff} \qquad (2.2.5)$$

where γ is a constant and H_{eff} represents the "effective" magnetic field (*). From a comparison of this equation with the second term of (2.1.24) and (2.1.25) we conclude that (**)

$$(H_{eff})_i = H_i - \frac{1}{\mu_0}\left[\frac{\partial F_1}{\partial \mathcal{M}_i} - \frac{1}{\rho}\frac{\partial}{\partial x_j}\left(\rho \frac{\partial F_1}{\partial \mathcal{M}_{i,j}}\right)\right]. \qquad (2.2.6)$$

Finally, if we put

$$a_{ji} = \frac{\partial F_1}{\partial \mathcal{M}_{i,j}} \qquad (2.2.7)$$

the second term with $d\mathcal{M}_i/dt$ as a factor will vanish. This determines the exchange tensor a_{ij}.

After collecting the remaining terms in Eq. (2.1.24) we arrive at the equation of heat conduction

$$Q_{i,i} = \rho r + \frac{\dot{j}^2}{\sigma} + \varkappa \underset{\sim}{j} \cdot \nabla T - \rho T \dot{S} \qquad (2.2.8)$$

(*) After multiplication of both sides of Eq. (2.2.5) by $\underset{\sim}{\mathcal{M}}$ we get $d\mathcal{M}^2/dt = 0$, and hence $\mathcal{M}^2 = \text{const}$. Eq. (2.2.5) therefore implies magnetic saturation.

(**) See [5], p. 84.

where expression (2.2.1) for the entropy has to be substituted.

To Eq. (2.2.8) the law of heat conduction has to be adjoined. If, for instance, Fourier's law is adopted in the form valid for a thermally isotropic body (*),

$$Q_i = -k\,T_{,i} + \varkappa T j_i$$

one obtains, after substitution into Eq. (2.2.8), assuming $k =$ const. and using $\nabla \cdot \underset{\sim}{j} = 0$ from Maxwell's equation (1.1.2) with $\rho_e = 0$,

(2.2.10) $$k \nabla^2 T = \rho T \dot{S} - \rho r - \frac{\dot{j}^2}{\sigma} + T \underset{\sim}{j} \cdot \nabla \varkappa .$$

The term $\dot{j}^2/\sigma$ represents the Joule heat production while the last term exhibits the Thomson effect. The coefficient $\varkappa$ will, in general, be temperature-dependent, $\nabla \varkappa = (d\varkappa/dT)\nabla T$.

Differential equation (2.2.2) has to be supplemented by boundary conditions. To this effect the Maxwell stress tensor m_{ij} is introduced as (**)

(2.2.11) $$m_{ji} = H_i B_j - \frac{1}{2}\mu_0 H^2 \delta_{ij}$$

(*) See [13] , § 25. An additional term appears in [13] which, however, is already included in Eq.(2.2.8).

(**) The Maxwell stress tensor is used here solely as an auxiliary quantity and no deeper meaning is ascribed to it.

and Eq. (2.2.2) is rewritten in the form

$$\left(\tau_{ji} + m_{ji}\right)_{,j} - \left(\frac{\partial \underset{\sim}{D}}{\partial t} \times \underset{\sim}{B}\right)_i = \rho \dot{v}_i . \qquad (2.2.12)$$

The displacement current $\partial \underset{\sim}{D}/dt$ will be neglected in the following.

Now, if V denotes the absolute velocity in the direction of its normal of a surface of discontinuity moving through the body, and if v_n is the velocity in the same direction of the corresponding body particle the jump condition (*)

$$\left[\tau_{ji} + m_{ji}\right] n_j = \left[\rho \left(v_n - V\right) v_i\right] \qquad (2.2.13)$$

follows from Eq. (2.2.12), where $[\varphi] := \overset{+}{\varphi} - \overset{-}{\varphi}$. To Eq. (2.2.13) the condition of continuity of mass has to be added,

$$\left[\rho \left(v_n - V\right)\right] = 0 . \qquad (2.2.14)$$

If the surface of discontinuity coincides with the surface of the body we have $V = v_n$ and $\overset{+}{\tau}_{ji} n_j = p_i$, where p_i is the external surface load. Remembering, furthermore, that the magnetic field intensity experiences a jump across the body surface (**) of magnitude

$$\overset{+}{H}_i - \overset{-}{H}_i = M_n n_i$$

(*) See, for instance, [11], p. 503 ff.

(**) See [5], p. 57.

while the normal component B_n of $\underset{\sim}{B}$ remains continuous, one obtains, utilizing Eq. (2.2.11),

$$[m_{ji}]n_j = B_n(\overset{+}{H}_i - \overset{-}{H}_i) - \frac{\mu_0}{2}(\overset{+}{H}_s - \overset{-}{H}_s)(\overset{+}{H}_s + \overset{-}{H}_s)\,n_i$$

$$= B_n M_n n_i - \frac{\mu_0}{2} M_n n_s(2\overset{+}{H}_s - M_n n_s)n_i = M_n(B_n - \mu_0 \overset{+}{H}_n + \frac{\mu_0}{2} M_n)n_i$$

(no summation over index n !).

But $B_n = \overset{+}{B}_n = \mu_0 \overset{+}{H}_n$. Hence, Eq. (2.2.13) finally renders the <u>boundary condition</u>

(2.2.16) $$\tau_{ji} n_j = p_i + \frac{\mu_0}{2} M_n^2 n_i .$$

In addition to body forces $\mu_0 M_j H_{j,i}$ and surface forces $\mu_0 M_n^2 n_i/2$ the magnetized body is also exposed to a distribution of couples as a consequence of the nonsymmetry of the stress tensor:

$$\tau_{ij} - \tau_{ji} = \rho\left(\frac{\partial F_1}{\partial \mathcal{M}_i}\mathcal{M}_j - \frac{\partial F_1}{\partial \mathcal{M}_j}\mathcal{M}_i + \frac{\partial F_1}{\partial \mathcal{M}_{i,A}}\mathcal{M}_{j,A} - \frac{\partial F_1}{\partial \mathcal{M}_{j,A}}\mathcal{M}_{i,A}\right).$$

(2.2.17)

This follows if one requires F_1 to remain invariant under an infinitesimal rigid rotation of the body, characterized by the antisymmetric tensor ω_{ij}. Under such a rotation, a vector v_i changes according to $dv_i = \omega_{ij} v_j$. Hence,

$$dF_1 = \frac{\partial F_1}{\partial x_{i,A}} dx_{i,A} + \frac{\partial F_1}{\partial \mathcal{M}_i} d\mathcal{M}_i + \frac{\partial F_1}{\partial \mathcal{M}_{i,A}} d\mathcal{M}_{i,A} =$$

$$= \left(\frac{\partial F_1}{\partial x_{i,A}} x_{j,A} + \frac{\partial F_1}{\partial \mathcal{M}_i} \mathcal{M}_j + \frac{\partial F_1}{\partial \mathcal{M}_{i,A}} \mathcal{M}_{j,A} \right) \omega_{ij} = 0 .$$

But, since ω_{ij} is antisymmetric, it follows that $a_{ji}\omega_{ij} = \frac{1}{2}(a_{ji}\omega_{ij} + a_{ij}\omega_{ji}) = \frac{1}{2}(a_{ji} - a_{ij})\omega_{ij}$, for an arbitrary tensor a_{ij}. Therefore,

$$\left(\frac{\partial F_1}{\partial x_{i,A}} x_{j,A} + \frac{\partial F_1}{\partial \mathcal{M}_i} \mathcal{M}_j + \frac{\partial F_1}{\partial \mathcal{M}_{i,A}} \mathcal{M}_{j,A} \right) -$$
$$- \left(\frac{\partial F_1}{\partial x_{j,A}} x_{i,A} + \frac{\partial F_1}{\partial \mathcal{M}_j} \mathcal{M}_i + \frac{\partial F}{\partial \mathcal{M}_{j,A}} \mathcal{M}_{i,A} \right) = 0 .$$

Using Eq. (2.2.4), one obtains Eq. (2.2.17).

If exchange forces are omitted, Eqs. (2.2.6) and (2.2.7) go over into

$$\frac{\partial F_1}{\partial \mathcal{M}_{i,A}} = 0 , \qquad H_i = \frac{1}{\mu_0} \frac{\partial F_1}{\partial \mathcal{M}_i} \tag{2.2.18}$$

and Eq. (2.2.17) reduces to

$$\tau_{ij} - \tau_{ji} = \mu_0 \left(H_i M_j - H_j M_i \right) . \tag{2.2.19}$$

In addition, using Eqs. $(2.1.18)_2$ and (2.2.11), one notes that now

$$\left(\tau_{ij} + m_{ij} \right) - \left(\tau_{ji} + m_{ji} \right) = 0 . \tag{2.2.20}$$

The sum of the Cauchy and the Maxwell stress tensor is now symmetric. Only for a magnetically linear material, $M_i = \chi_m H_i$, the Cauchy tensor itself is symmetric.

Integration of Eqs. (2.2.2) and (2.2.17) over the entire volume of the body leads to very simple expressions for the resultant force and moment, [5], p. 86. One finds for $\dot{\underset{\sim}{j}} = \underset{\sim}{0}$,

$$\underset{\sim}{F} = \mu_0 \int_V (\underset{\sim}{M} \cdot \nabla) \underset{\sim}{H} \, dV + \frac{\mu_0}{2} \oint_{\partial V} M_n^2 \underset{\sim}{n} \, d\partial V = \mu_0 \int_V (\underset{\sim}{M} \cdot \nabla) \underset{\sim}{H}_0 \, dV \tag{2.2.21}$$

for the force, and

$$\underset{\sim}{L} = \mu_0 \left[\int_V \underset{\sim}{r} \times (\underset{\sim}{M} \cdot \nabla) \underset{\sim}{H} \, dV + \frac{1}{2} \oint_{\partial V} \underset{\sim}{r} \times \underset{\sim}{n} M_n^2 \, d\partial V + \int_V \underset{\sim}{M} \times \underset{\sim}{H} \, dV \right] =$$

$$= \mu_0 \left[\int_V \underset{\sim}{r} \times (\underset{\sim}{M} \cdot \nabla) \underset{\sim}{H}_0 \, dV + \int \underset{\sim}{M} \times \underset{\sim}{H}_0 \, dV \right] \tag{2.2.22}$$

for the moment. $\underset{\sim}{H}_0$ is the applied field, produced by the entire surrounding.

Chapter 3
THERMOMECHANICS OF DIELECTRIC BODIES

We consider an elastic, polarized but not magnetized medium which is an electrical insulator but may conduct heat, i.e., we have now $\underset{\sim}{j} = 0$ and $\underset{\sim}{M} = 0$, but $\underset{\sim}{P} \neq 0$.

3.1 The Quasistatic Case

In this case, and in the absence of an electric charge, the Maxwell equations (1.1.1) reduce, if $\partial\underset{\sim}{D}/\partial t$ is neglected, to

$$\left.\begin{array}{ll} \nabla \times \underset{\sim}{H} = 0, & \nabla \times \underset{\sim}{E} = -\dfrac{\partial \underset{\sim}{B}}{\partial t} \\ \nabla \cdot \underset{\sim}{D} = 0, & \nabla \cdot \underset{\sim}{B} = 0 \end{array}\right\} \tag{3.1.1}$$

while Eqs. (1.2.10) read

$$\underset{\sim}{D} = \varepsilon_0 \underset{\sim}{E} + \underset{\sim}{P} \qquad \underset{\sim}{B} = \mu_0 \underset{\sim}{H} . \tag{3.1.2}$$

The term $\underset{\sim}{v} \times \underset{\sim}{P}$ is omitted. Furthermore, in contradistinction to the ferromagnetic body where the magnetization gradient is of great importance, the influence of the polarization gradient is neglected (*), i.e., the free energy is assumed in the

---

(*) Mindlin [16] and his coworkers have investigated this influence.

form (*)

(3.1.3) $$F = F_2(x_{i,A}, \mathcal{P}_i, T)$$

where

(3.1.4) $$\mathcal{P}_i = \frac{P_i}{\rho}$$

represents polarization per unit of mass. The energy balance reads then

$$\frac{d}{dt}\int_V \left[\rho\left(U + \frac{v^2}{2}\right) + U_e\right] dV = \int_V (\rho r + f_i v_i)\, dV +$$

(3.1.5) $$\oint_{\partial V} \left[\overset{*}{t}_{ij} v_j - Q_i - (\underset{\sim}{E} \times \underset{\sim}{H})_i + U_e v_i\right] n_i\, d\partial V .$$

The derivation of the basic equations now follows precisely the same line as in the preceding chapter with polarization P_i playing the rôle of magnetization M_i. The expression (2.1.16) for the electromagnetic energy is retained without change and renders

$$\frac{\partial U_e}{\partial t} = -\nabla\cdot(\underset{\sim}{E} \times \underset{\sim}{H}) - E_i \frac{\partial P_i}{\partial t} = -\nabla\cdot(\underset{\sim}{E} \times \underset{\sim}{H}) -$$

$$- \rho E_i \dot{\mathcal{P}}_i + (\rho E_i \mathcal{P}_i v_k)_{,k} - \rho \mathcal{P}_i v_k E_{i,k} .$$

(*) Instead of F_2, <u>Tiersten</u> [7] employs the function $\chi(x_{i,A}, E_i, T) = F_2 - E_i \mathcal{P}_i$. Eq. (3.1.7) is then to be replaced by $\mathcal{P}_i = -\partial\chi/\partial E_i$.

One obtains the following set of equations:

$$\rho \dot{v}_i = f_i + \tau_{ji,j} + P_k E_{k,i} \tag{3.1.6}$$

$$S = -\frac{\partial F_2}{\partial T}, \quad E_i = \frac{\partial F_2}{\partial \mathcal{P}_i} \tag{3.1.7}$$

$$\tau^*_{ij} = \tau_{ij} + E_s P_s \delta_{ij} \tag{3.1.8}$$

$$Q_{i,i} + \rho T \dot{S} - \rho r = 0 . \tag{3.1.9}$$

The boundary condition for Eq. (3.1.6) is analogous to Eq. (2.2.16),

$$\tau_{ji} n_j = p_i + \frac{\varepsilon_0}{2} P_n^2 n_i . \tag{3.1.10}$$

The stress tensor τ_{ij} is again defined by Eq. (2.2.4), with F_1 replaced by F_2 . Similarly, we have

$$\tau_{ij} - \tau_{ji} = E_i P_j - E_j P_i . \tag{3.1.11}$$

In the static case Eq. $(3.1.1)_2$ goes over into $\nabla \times \underset{\sim}{E} = 0$, i.e., $E_{k,i} = E_{i,k}$.

3.2 The General Case

A theory for the general, nonstationary case was given by Toupin [4]. Starting with the Lorentz force density on an electric charge and current (*)

(3.2.1) $$\underset{\sim}{f}_L = \rho_e \underset{\sim}{E} + \underset{\sim}{j} \times \underset{\sim}{B}$$

we use, for the polarized body, Eqs. $(1.1.8)_3$ and (1.2.9b), and replace ρ_e by the density $-\nabla \cdot \underset{\sim}{P}$ and $\underset{\sim}{j}$ by the current $\partial \underset{\sim}{P}/\partial t + $ $+ \nabla \times (\underset{\sim}{P} \times \underset{\sim}{v})$, to obtain

(3.2.2a) $$\underset{\sim}{f}_L = -\underset{\sim}{E}(\nabla \cdot \underset{\sim}{P}) + \left[\frac{\partial \underset{\sim}{P}}{\partial t} + \nabla \times (\underset{\sim}{P} \times \underset{\sim}{v})\right] \times \underset{\sim}{B}$$

or, equivalently,

(3.2.2b) $$\underset{\sim}{f}_L = -(\underset{\sim}{E} + \underset{\sim}{v} \times \underset{\sim}{B})(\nabla \cdot \underset{\sim}{P}) + \frac{d_c \underset{\sim}{P}}{dt} \times \underset{\sim}{B}\,.$$

With this assumption, we may now write the equation of motion as

(3.2.3) $$\rho \dot{v}_i = t_{ji,j} + f_i - E'_i(\nabla \cdot \underset{\sim}{P}) + (\underset{\sim}{P}^* \times \underset{\sim}{B})_i$$

where

(3.2.4) $$\underset{\sim}{E}' = \underset{\sim}{E} + \underset{\sim}{v} \times \underset{\sim}{B}, \quad \underset{\sim}{P}^* = \frac{d_c \underset{\sim}{P}}{dt}\,.$$

The stress tensor t_{ij} is not identical with the stress tensor τ_{ij} of Eq. (3.1.6). The tensor t_{ij} is symmetric.

(*) See [10], p. 96.

The law of energy balance is taken in the form

$$\rho(\dot{U} - r) = t_{ji} v_{i,j} + P_i^* E_i' - Q_{i,i} . \tag{3.2.5}$$

This postulate may be made plausible as follows. From Eq. (1.3.2) we note that the quantity $\underset{\sim}{j} \cdot \underset{\sim}{E}$ represents the rate at which the material gains energy from the electromagnetic field. Subtracting from this the work done by the Lorentz force, the usual expression for the rate of change of internal energy (*) has to be changed to read

$$\rho\dot{U} = t_{ji} v_{i,j} + \rho r - Q_{i,i} + \underset{\sim}{j} \cdot \underset{\sim}{E} - \underset{\sim}{f}_L \cdot \underset{\sim}{v} . \tag{3.2.6}$$

Substituting for $\underset{\sim}{j}$ the expression $\underset{\sim}{P}^* - (\nabla \cdot \underset{\sim}{P}) \underset{\sim}{v}$, and for f_L Eq. (3.2.2b), one obtains Eq. (3.2.5).

A material measure of polarization $\underset{\sim}{\Pi}$ is now defined through

$$P_i = \frac{\rho}{\rho_0} x_{i,A} (\Pi_A) \tag{3.2.7}$$

where ρ_0 is the mass density in the initial undeformed state,

$$\frac{\rho_0}{\rho} = \det \left| x_{i,A} \right| . \tag{3.2.8}$$

(*) See, for instance, [17] , p. 78.

Similar to Eq. (3.1.3) the free energy is introduced as

$$F = F_3(x_{i,A}, \Pi_A, T). \tag{3.2.9}$$

From this we have

$$\rho\dot{U} = \rho\Big(\frac{\partial F_3}{\partial x_{i,A}} v_{i,A} + \frac{\partial F_3}{\partial \Pi_A}\dot{\Pi}_A + \frac{\partial F_3}{\partial T}\dot{T} + S\dot{T} + T\dot{S}\Big). \tag{3.2.10}$$

The polarization satisfies the identity

$$P_i^* = \frac{\rho}{\rho_0} x_{i,A}\dot{\Pi}_A . \tag{3.2.11}$$

Proof (*)

Inverting Eq. (3.2.7) and differentiating we obtain, upon putting $\rho_0/\rho = \mathfrak{J}$,

$$\dot{\Pi}_A = \dot{\mathfrak{J}} X_{A,i} P_i + \mathfrak{J}\dot{X}_{A,i} P_i + \mathfrak{J} X_{A,i}\dot{P}_i .$$

Differentiating $\mathfrak{J}\rho = \rho_0$ and using Eq. (2.1.20) in the form $\dot{\rho} + \rho v_{i,i} = 0$, we find

$$\dot{\mathfrak{J}} = \mathfrak{J} v_{i,i} .$$

Hence,

$$\frac{1}{\mathfrak{J}}\dot{\Pi}_A = X_{A,i}P_i v_{j,j} + \dot{X}_{A,i}P_i + X_{A,i}\dot{P}_i = \dot{X}_{A,i}P_i + X_{A,i}(P_i^* + P_j v_{i,j}) .$$

(*) This proof was communicated to the author by Prof. M.E. Gurtin

But differentiation of the relation $X_{A,i}\, x_{i,B} = \delta_{AB}$ yields

$$\dot{X}_{A,i}\, x_{i,B} + X_{A,i}\, v_{i,B} = 0$$

whence

$$\dot{X}_{A,i} = - X_{A,j}\, v_{j,i} \, .$$

If this is substituted in the equation above, Eq. (3.2.11) is obtained.
If Eqs. (3.2.10) and (3.2.11) are substituted into Eq. (3.2.5) one obtains, with $v_{i,A} = v_{i,j}\, x_{j,A}$, after comparing coefficients,

$$t_{ji} = \rho\, x_{j,A} \frac{\partial F_3}{\partial x_{i,A}} \tag{3.2.12}$$

$$E'_i = \rho_0 X_{A,i} \frac{\partial F_3}{\partial \Pi_A} \tag{3.2.13}$$

and

$$Q_{i,i} = \rho(r - T\dot{S}), \qquad S = - \frac{\partial F_3}{\partial T} \, . \tag{3.2.14}$$

It should be noted that the vector E'_i is not uniquely determined by Eq. (3.2.5). This equation contains only the product $E'_i P^*_i$. Therefore, the vector $\underset{\sim}{E}' + \underset{\sim}{G} \times \underset{\sim}{P}^*$ would also represent a solution, where $\underset{\sim}{G}$ is an arbitrary vector. For details, see [4] , where also applications to predeformed bodies may be found.

Chapter 4

WAVE PROPAGATION

Most applications in thermomagneto- and thermoelectroelasticity are concerned with wave propagation phenomena. However, these investigations are, in general, not based on the general equations of the preceding chapters but make use of simplified, linearized versions which will be given below.

Kaliski and Nowacki, in a long series of publications (*), beginning 1959, have treated, under various aspects, the problem of infinitesimal waves in elastic and thermoelastic media with electromagnetic effects included. Both linear magnetic bodies with perfect and with finite electric conductivity as well as linear dielectrics are discussed. In the linearized version of equation (2.2.10) of heat conduction, an additional term is introduced which accounts for an "inertia effect" in the propagation of heat:

$$Q_{i,i} + \tau \dot{Q}_{i,i} = -(k_{ij} T_{,j})_{,i} + (\Pi_{ik} j_k)_{,i} .$$

Willson [19], by extending and correcting a preceding paper by Paria (**) has treated the propagation of a plane

(*) A list of corresponding references may be found in [18].

(**) cf. the survey article [20] by Paria.

infinitesimal magneto-thermoelastic wave in a homogeneous and isotropic medium, exposed to an applied ("primary") magnetic field. Magneto-thermoelastic acceleration waves in a prestressed, initially isotropic medium have been discussed by McCarthy [21].

4.1. Linearized Magneto - Thermoelastic Equations

As a consequence of the assumption of small deformations, the linearized strain- displacement relations

$$2\varepsilon_{ij} = u_{i,j} + u_{j,i} \tag{4.1.1}$$

are introduced, where $u_i = x_i - X_i$ is the displacement vector. Furthermore, the various stress tensors become identical and will be denoted by σ_{ij}. Also, the distinction between $\partial/\partial t$ and d/dt disappears in the equation of motion. A dot will now denote $\partial/\partial t$.

Further assumptions are (*):

a) Quasistationary electromagnetic field. Maxwell's Eqs. (1.1.1) reduce then, if charges are absent, to

$$\left.\begin{array}{ll} \nabla \times \underset{\sim}{H} = \underset{\sim}{j}, & \nabla \times \underset{\sim}{E} = -\dot{\underset{\sim}{B}} \\ \nabla \cdot \underset{\sim}{D} = 0 & \nabla \cdot \underset{\sim}{B} = 0 \,. \end{array}\right\} \tag{4.1.2}$$

b) The equation of motion is simplified to read

$$\rho_0 \ddot{u}_i = \sigma_{ji,j} + (\underset{\sim}{j} \times \underset{\sim}{B})_i \tag{4.1.3}$$

where σ_{ij} is the usual stress tensor for small deformations.

(*) See [20], p. 93.

c) Hooke's law is introduced for Eq. (2.2.4), reading, for an isotropic body,

(4.1.4) $$\sigma_{ij} = 2G\varepsilon_{ij} + (\lambda e - \beta\vartheta)\delta_{ij}$$

where $e = \varepsilon_{ii}$. The constants G and λ are those of Lamé, and

(4.1.5) $$\beta = \frac{E\alpha_T}{1-2\nu} = (3\lambda + 2G)\alpha_T$$

where E is Young's modulus, ν is Poisson's ratio and α_T the coefficient of thermal expansion. Piezoelectric effects are, in an isotropic body, of second order.

d) As a consequence of Eq. (4.1.4), Eq. (2.2.10) reduces to the usual equation of heat conduction of thermoelasticity (*),

(4.1.6) $$k\nabla^2\vartheta = \rho_0 c\dot{\vartheta} + \beta T_0 \dot{e}$$

where $\vartheta = T - T_0$ and c is the specific heat. T_0 is absolute temperature and ρ_0 is the mass density in the undeformed body. Heat sources are assumed to be absent, $r = 0$.

e) Eq. (3.2.6) for the rate of change of internal energy reads now, with $\underset{\sim}{f}_L = \underset{\sim}{j} \times \underset{\sim}{B}$,

(4.1.7) $$\rho_0\dot{U} = \sigma_{ij}\dot{\varepsilon}_{ji} + \left[E_i + (\dot{\underset{\sim}{u}} \times \underset{\sim}{B})_i\right] \cdot j_i - Q_{i,i} \,.$$

(*) See [17] , p. 84.

f) Eqs. (1.2.12) are used rather than Eqs. (1.2.13),

$$D_i = \varepsilon E_i, \quad B_i = \mu H_i . \tag{4.1.8}$$

4.2 Infinitesimal Magneto - Thermoelastic Waves

We follow Willson [19] and assume the body to be exposed to an applied magnetic field

$$\underset{\sim}{B}_0 = \begin{pmatrix} B_1 \\ B_2 \\ B_3 \end{pmatrix} \tag{4.2.1}$$

constant in space and time. An electromagnetic and a displacement plane wave are supposed to travel through the body in the direction of $x_1 = x$. All quantities are assumed to be functions of x and t only.

We take the magnetic field in the form

$$\underset{\sim}{B} = \underset{\sim}{B}_0 + \underset{\sim}{b} \tag{4.2.2}$$

where $\underset{\sim}{b}$ is a small perturbation. Together with the other plane wave perturbations it is introduced by putting

$$\begin{aligned} \underset{\sim}{u}^T(x,t) &= (u, v, w)\exp\left[i(\gamma x - \omega t)\right] \\ \underset{\sim}{b}^T(x,t) &= (b_1, b_2, b_3)\exp\left[i(\gamma x - \omega t)\right] \end{aligned} \tag{4.2.3}$$

$$\underset{\sim}{j}^T(x,t) = (j_1, j_2, j_3)\exp\left[i(\gamma x - \omega t)\right]$$

(4.2.3) $$\underset{\sim}{E}^T(x,t) = (E_1, E_2, E_3)\exp\left[i(\gamma x - \omega t)\right]$$

$$\vartheta(x,t) = \tau \exp\left[i(\gamma x - \omega t)\right] .$$

The amplitudes are constant. The factor $\exp\left[i(\gamma x - \omega t)\right]$ will be suppressed from here on.

From Eq. (4.1.1) we obtain for the strain

(4.2.4) $$(\varepsilon_{ij}) = \frac{i\gamma}{2}\begin{pmatrix} 2u & v & w \\ v & 0 & 0 \\ w & 0 & 0 \end{pmatrix}$$

and from Eq. (4.1.4) for the stress

(4.2.5) $$(\sigma_{ij}) = \begin{pmatrix} (\lambda + 2G)i\gamma u - \beta\tau & Gi\gamma v & Gi\gamma w \\ Gi\gamma v & \lambda i\gamma u - \beta\tau & 0 \\ Gi\gamma w & 0 & \lambda i\gamma u - \beta\tau \end{pmatrix} .$$

Using $e = i\gamma u$ from Eq. (4.2.4), one obtains from the equation of heat conduction, Eq. (4.1.6),

$$\tau = \alpha u, \qquad \alpha = \frac{\beta T_0 \gamma\omega}{\rho_0 c i\omega - k\gamma^2} .$$

The equation of motion (4.1.3) renders, with $\underset{\sim}{j} \times \underset{\sim}{B} =$

$= \underset{\sim}{j} \times \underset{\sim}{B}_0 + \ldots ,$

$$-\rho_0\omega^2(u, v, w) = i\gamma\Big(\big[(\lambda + 2G)i\gamma - \alpha\beta\big]u,\quad Gi\gamma v,\quad Gi\gamma w\Big)$$
$$+(j_2B_3 - j_3B_2,\quad j_3B_1 - j_1B_3,\quad j_1B_2 - j_2B_1)\,. \tag{4.2.7}$$

We now use Eqs. (4.1.2) to determine the magnetic field in the body. First, since $\partial/\partial y = \partial/\partial z = 0$, it follows from

$$\nabla\cdot\underset{\sim}{B} = \nabla\cdot\underset{\sim}{b} = \frac{\partial}{\partial x}\Big(b_1\exp\big[i(\gamma x - \omega t)\big]\Big) = 0,$$

that $b_1 = 0$. Then

$$\nabla\times\underset{\sim}{E} = i\gamma(0, -E_3, E_2) = i\omega(0, b_2, b_3)$$

whence

$$E_2 = \frac{\omega}{\gamma}b_3\,,\qquad E_3 = -\frac{\omega}{\gamma}b_2\,. \tag{4.2.8}$$

The first of Eqs. (4.1.2) gives, if Eq. $(4.1.8)_2$ is substituted,

$$i\gamma(0, -b_3, b_2) = \mu(j_1, j_2, j_3)$$

and, hence,

$$j_1 = 0,\qquad j_2 = -\frac{i\gamma}{\mu}b_3,\qquad j_3 = \frac{i\gamma}{\mu}b_2\,. \tag{4.2.9}$$

Ohm's law, Eq. (2.1.22),

$$\underset{\sim}{j} = \sigma(\underset{\sim}{E} + \dot{\underset{\sim}{u}}\times\underset{\sim}{B} - \varkappa\,\nabla\vartheta) \tag{4.2.10}$$

renders, with $\dot{\underset{\sim}{u}} \times \underset{\sim}{B} = \dot{\underset{\sim}{u}} \times \underset{\sim}{B}_0 + \ldots$ and $\nabla\vartheta = i\gamma(\tau,0,0)$,

$$\frac{i\gamma}{\mu}(0,-b_3,b_2) = \sigma\Big[E_1 - i\omega(vB_3 - wB_2) - i\gamma\chi\tau,$$

(4.2.11)

$$\frac{\omega}{\gamma}b_3 - i\omega(wB_1 - uB_3), -\frac{\omega}{\gamma}b_2 - i\omega(uB_2 - vB_1)\Big].$$

The x-component of this equation determines E_1 and will not be considered here any further. For the remaining five unknown quantities u, v, w, b_2, b_3 we have the following five equations:
from Eqs. (4.2.7) and (4.2.9)

$$\left.\begin{aligned}
&\left[-\rho_0\omega^2 + (\lambda+2G)\gamma^2 + i\alpha\beta\gamma\right]u + \frac{i\gamma}{\mu}(B_2b_2 + B_3b_3) &&= 0\\
&(-\rho_0\omega^2 + G\gamma^2)v - \frac{i\gamma}{\mu}B_1b_2 &&= 0\\
&(-\rho_0\omega^2 + G\gamma^2)w - \frac{i\gamma}{\mu}B_1b_3 &&= 0\\
&\text{from Eq. (4.2.11)}\\
&(i\sigma\omega B_3)u - (i\sigma\omega B_1)w + \left(\frac{i\gamma}{\mu} + \frac{\sigma\omega}{\gamma}\right)b_3 &&= 0\\
&(i\sigma\omega B_2)u - (i\sigma\omega B_1)v + \left(\frac{i\gamma}{\mu} + \frac{\sigma\omega}{\gamma}\right)b_2 &&= 0\,.
\end{aligned}\right\}$$

(4.2.12)

We now choose our y and z axes so as to make $b_3 = 0$. It then follows from $(4.2.12)_3$ that, unless $\rho_0\omega^2 = G\gamma^2$, the displacement $w = 0$ and, from $(4.2.12)_4$ that, if $w = 0, B_3$ must be zero. Thus, if we put

$$w = 0, \quad B_3 = 0, \quad b_3 = 0 \tag{4.2.13}$$

Eqs. $(4.2.12)_{3,4}$ are satisfied and we are left with three equations in the three unknowns u, v, b_2. For a non-trivial solution the determinant of the system must vanish:

$$\begin{vmatrix} -\rho_0\omega^2+(\lambda+2G)\gamma^2+i\alpha\beta\gamma & 0 & \frac{i\gamma}{\mu}B_2 \\ 0 & -\rho_0\omega^2+G\gamma^2 & -\frac{i\gamma}{\mu}B_1 \\ i\sigma\omega B_2 & -i\sigma\omega B_1 & \frac{i\gamma}{\mu}+\frac{\sigma\omega}{\gamma} \end{vmatrix} = 0 .$$

Before proceding, we introduce some abbreviations:

$$\left.\begin{aligned} &V_L = \sqrt{\frac{\lambda+2G}{\rho_0}}, \quad V_T = \sqrt{\frac{G}{\rho_0}}, \quad \omega^* = \frac{\rho_0 c}{k}V_L^2, \quad \nu = \frac{\omega}{\omega^*} \\ &\xi = \frac{\gamma V_L}{\omega^*}, \quad \varepsilon_E = \frac{\rho_0 c}{k\sigma\mu}, \quad \varepsilon_T = \frac{\beta^2 T_0}{\rho_0^2 c V_L^2} . \end{aligned}\right\} \tag{4.2.15}$$

Here, V_L is the speed of the longitudinal, and V_T the speed of the transverse elastic wave. ω^* is the "characteristic frequency". ε_E and ε_T are the coefficients of electromagnetic and thermal coupling, respectively. Eq. (4.2.14) now reads

$$\left[\left(\frac{V_T}{V_L}\xi\right)^2 - \nu^2\right]\left[\left\{(\xi^2-\nu^2)(\nu+i\varepsilon_E\xi^2) + \frac{B_2^2}{\rho_0\mu V_L^2}\xi^2\nu\right\}(\nu+i\xi^2) + \right.$$

$$\left. + \varepsilon_T\xi^2\nu(\nu+i\varepsilon_E\xi^2)\right] + \frac{B_1^2}{\rho_0\mu V_L^2}\xi^2\nu\left[(\xi^2-\nu^2)(\nu+i\xi^2)+\varepsilon_T\xi^2\nu\right] = 0 .$$

(4.2.16)

Several cases will now be considered.

Case (a) Zero applied field, $B_1 = B_2 = 0$.

The only surviving terms in the determinant are along its principal diagonal. The roots are:

(I) $\gamma^2 = i\omega\sigma\mu$ or $\nu + i\varepsilon_E\xi^2 = 0$. This corresponds to $u = v = w = 0$. The electromagnetic wave is not coupled to the thermal and elastic field.

(II) $\rho_0\omega^2 = G\gamma^2$ or $\nu = \xi V_T/V_L$. This corresponds to $\underset{\sim}{u}^T = (0, v, 0)$ giving a transverse elastic wave not coupled either to the thermal or the electromagnetic field.

(III) $-\rho_0\omega^2 + (\lambda + 2G)\gamma^2 + i\alpha\beta\gamma = 0$ or $(\xi^2-\nu^2)(\nu+i\xi^2)+\varepsilon_T\xi^2\nu = 0$. This corresponds to $\underset{\sim}{u}^T = (u, 0, 0)$ giving a longitudinal attenuated thermoelastic wave, uncoupled to the electromagnetic field, cf. [17], p. 99.

Case (b) $B_1 = 0, B_2 \neq 0$. This corresponds to a purely transverse magnetic field. Eq. (4.2.16) factors into two parts. For $\xi V_T = \nu V_L$ we have Case (aII) of an uncoupled transverse elastic wave, while

$$\left\{(\xi^2-\nu^2)(\nu+i\varepsilon_E\xi^2)+\frac{B_2^2}{\rho_0\mu V_L^2}\xi^2\nu\right\}(\nu+i\xi^2)+\varepsilon_T\xi^2\nu(\nu+i\varepsilon_E\xi^2)=0$$

(4.2.17)

represents an electromagnetic wave coupled with a longitudinal thermoelastic wave.

Assuming a perfect electric conductor, $\sigma\to\infty$, i.e., putting $\varepsilon_E = 0$, Eq. (4.2.17) goes over into

$$\left[\left(1+\frac{B_2^2}{\rho_0\mu V_L^2}\right)\xi^2-\nu^2\right](\nu+i\xi^2)+\varepsilon_T\xi^2\nu=0. \qquad (4.2.18)$$

In comparison with the case $B_2 = 0$ it follows that the effect of the transverse magnetic field is to increase the speed of propagation V_L of the longitudinal isothermal elastic wave by the factor $(1+B_2^2/\rho_0\mu V_L^2)^{1/2}$. This may be seen by putting $\varepsilon_T = 0$ in Eq. (4.2.18). Considering then waves of an assigned frequency, i.e., regarding ν as a fixed real constant, the resulting two roots $\xi_{1,2}$ determine the speed V_1 of the displacement wave and V_2 of the temperature wave as

$$V_{1,2}=\frac{\omega}{\mathrm{Re}\,\gamma_{1,2}}=V_L\frac{\nu}{\mathrm{Re}\,\xi_{1,2}}$$

cf. Eq. (4.2.3). But

$$\frac{\nu}{\mathrm{Re}\,\xi_1} = \frac{\nu}{\xi_1} = \left(1 + \frac{B_2^2}{\rho_0 \mu V_L^2}\right)^{1/2}$$

for the displacement wave, and

$$\frac{\nu}{\mathrm{Re}\,\xi_2} = \frac{\nu}{\mathrm{Re}\sqrt{i\nu}} = \sqrt{2\nu}$$

for the thermal wave. The latter is attenuated but not influenced by the magnetic field.

Case (c) If both B_1 and B_2 are different from zero, i.e., if the applied magnetic field has both transverse and longitudinal components, the transverse and longitudinal elastic waves are linked together.

4.3 Magneto - Thermoelastic Acceleration Waves

A general treatment of these waves has been given by McCarthy [21]. We shall be satisfied here with a simplified version, assuming, in addition to the linearized equations of Sec. 4.1., a perfect electrical conductor, $1/\sigma = 0$. With Eq. (4.2.10), this assumption renders

$$(4.3.1) \qquad E_i = -(\dot{\underset{\sim}{u}} \times \underset{\sim}{B})_i + \varkappa\, \vartheta_{,i} = e_{ijk} \dot{u}_k B_j + \varkappa\, \vartheta_{,i}$$

where e_{ijk} is the permutation tensor (*).

In an acceleration wave the following conditions hold at the wave front:

(a) The quantities u_i, $\dot{u}_i$, ε_{ij}, ϑ, E_i and B_i as well as their tangential derivatives are continuous functions of x_i and t. We do not explicitly separate B_i here into the constant applied field and the small perturbation.

(b) The acceleration $\ddot{u}_i$ suffers a jump $[\ddot{u}_i] = \ddot{u}^+ - \ddot{u}^-$ across the wave front which will be assumed to be the plane $x_1 \equiv x = \xi(t)$. Then $\dot{\xi} = V$ is the speed of propagation of the wave. All quantities are assumed independent of $x_2 \equiv y$ and $x_3 \equiv z$.

With these assumptions one obtains from the equation of motion (4.1.3), taking jumps,

$$\rho_0[\ddot{u}_i] = [\sigma_{xi,x}] + e_{ijk}[j_j]B_k . \qquad (4.3.2)$$

Now, for a function F, continuous with continuous derivatives,

---

(*) If the temperature gradient is neglected or absent, an interesting consequence results from Eq. (4.3.1). This may be seen by rewriting Eq. $(1.1.1)_2$ in the form $\nabla \times (\underset{\sim}{E} + \underset{\sim}{v} \times \underset{\sim}{B}) + d_c \underset{\sim}{B}/dt = 0$, where $d_c \underset{\sim}{B}/dt$ is the convected time flux as defined by Eq. (1.2.3a). It follows that $d_c \underset{\sim}{B}/dt = 0$ for a perfect conductor. Thus, the magnetic induction, measured relative to the body, remains fixed during the motion of the material. In other words, the lines of magnetic induction are "frozen" to the particles of the body, a phenomenon well-known in magnetohydrodynamics.

the following jump relation holds (*)

(4.3.3) $$\left[\frac{\partial^2 F}{\partial t^2}\right] = V^2\left[\frac{\partial^2 F}{\partial x^2}\right] = -V\left[\frac{\partial^2 F}{\partial x \partial t}\right].$$

Hence, we may write,

(4.3.4) $$\left[\ddot{u}_i\right] = V^2 a_i\,, \qquad a_i = \left[u_{i,xx}\right] = -\frac{1}{V}\left[\dot{u}_{i,x}\right]$$

where a_i is the amplitude vector of the wave.

In view of Eq. (4.3.1), the energy equation (4.1.7) reduces to

(4.3.5a) $$\rho_0 \dot{U} = \sigma_{ij}\dot{u}_{j,i} - Q_{i,i} + \varkappa j_i \vartheta_{,i}\,.$$

In integral form, this equation reads (*)

(4.3.5b) $$\frac{d}{dt}\int_m U\,dm = \int_V (\sigma_{ij}\dot{u}_{j,i} + \varkappa j_i \vartheta_{,i})\,dV - \oint_{\partial V} Q_i n_i\,d\partial V\,.$$

Now, consider the volume between two plane surfaces of unit area fixed in space at $x_1 = \xi(t) - \lambda$ and $x_2 = \xi(t) + \lambda$. If Eq. (4.3.5b) is taken over this volume we get, in the limit $\lambda \to 0$,

$$\lim_{\lambda\to 0}\left\{\frac{d}{dt}\int_{x_1}^{\xi(t)} U\,dm + \frac{d}{dt}\int_{\xi(t)}^{x_2} U\,dm - \int_{x_1}^{x_2}\sigma_{xj}\dot{u}_{j,x}\,dx - \varkappa\int_{x_1}^{x_2} j_x \vartheta_{,x}\,dx\right\} = -\left[Q_x\right].$$

(*) cf. [17].

Performing the differentiation, we obtain for the left-hand side,

$$\lim_{\lambda\to 0}\left\{(V\rho_0 U)_1 - (V\rho_0 U)_2\right\} = -V\rho_0\left[U\right]$$

and, since σ_{xj} is continuous,

$$\lim_{\lambda\to 0}\int_{x_1}^{x_2}\sigma_{xj}\,\dot{u}_{j,x}\,dx = \sigma_{xj}\left[\dot{u}_j\right]. \qquad (4.3.6)$$

Assuming here U to depend on strain, temperature and magnetization, which are continuous, U is continuous. Also, $\dot{u}_j$ is continuous. Finally, acceleration waves are <u>homothermal</u>, i.e., $\dot{\vartheta}$ and $\vartheta_{,x}$ are continuous across the wave front, provided the heat conduction modulus $-\partial q_i/\partial\vartheta_{,j}$ is positive definite (*). Hence,

$$\lim_{\lambda\to 0}\int_{x_1}^{x_2} j_x\,\vartheta_{,x}\,dx = 0$$

and it follows that the heat flux is continuous,

$$\left[Q_x\right] = 0\,. \qquad (4.3.7)$$

In order to find the jump $\left[j_i\right]$ we turn to the discontinuity relations (1.1.5), where the term $V\left[D\right]$ has to be

---

(*) See 12 , p. 384.

omitted as a consequence of Eq. $(4.1.2)_1$. Since $\underset{\sim}{E}$ and $\underset{\sim}{H}$ are continuous, we differentiate with respect to x and obtain, using the permutation tensor,

$$(4.3.8) \qquad \begin{cases} e_{ijk} n_j \left[H_{k,x}\right] = \left[j_i\right] \\ e_{ijk} n_j \left[E_{k,x}\right] = V\left[B_{i,x}\right] \end{cases}$$

where $n_x = 1,\ n_y = n_z = 0$. From Eq. (4.3.1) we have

$$E_{i,x} = e_{ijk}(\dot{u}_{k,x} B_j + \dot{u}_k B_{j,x}) + \varkappa\, \vartheta_{,ix}$$

and, on taking jumps and using Eq. $(4.3.4)_2$,

$$\left[E_{i,x}\right] = e_{ijk}\left(\dot{u}_k \left[B_{j,x}\right] - V a_k B_j\right) + \varkappa \left[\vartheta_{,ix}\right] .$$

The product $\dot{u}_k\left[B_{j,x}\right]$ has to be neglected in a linearized theory. Furthermore, from Eq. (4.1.6) with $\dot{\vartheta}$ continuous,

$$k\left[\vartheta_{,xx}\right] = \beta T_0 \left[\dot{u}_{x,x}\right] = -\beta T_0 V a_x .$$

Hence,

$$(4.3.9) \qquad \left[E_{i,x}\right] = -V\left(e_{ijk}\, a_k B_j + \varsigma\, a_x\, \delta_{ix}\right)$$

where

$$(4.3.10) \qquad \varsigma = \frac{\varkappa}{k}\, \beta T_0 .$$

If Eq. (4.3.9) is substituted into Eq. $(4.3.8)_2$, we get, using Eq. (4.1.8),

$$\left[B_{i,x}\right] = \mu\left[H_{i,x}\right] = -e_{ijk}\,n_j\,(e_{klm}\,B_l a_m + \xi a_x \delta_{kx}). \qquad (4.3.11)$$

Eq. $(4.3.8)_1$ now renders for the desired jump $\left[\dot{j}_i\right]$

$$\left[\dot{j}_i\right] = -\frac{1}{\mu}\,e_{ijk}\,e_{klm}\,n_j\,n_l\,(e_{mrs}\,B_r a_s + \xi a_x \delta_{mx}). \qquad (4.3.12)$$

To conclude the set of jump equations, we use Hooke's law (4.1.4) and obtain, after differentiation and summation,

$$\sigma_{xi,x} = 2G\,\varepsilon_{xi,x} + (\lambda e_{,x} - \beta\vartheta_{,x})\,\delta_{ix}$$

$$= G u_{i,xx} + (G+\lambda)\,u_{x,xx}\delta_{ix} - \beta\vartheta_{,x}\delta_{ix}\,.$$

Taking jumps and utilizing Eq. (4.3.4), we find,

$$\left[\sigma_{xi,x}\right] = G\,a_i + (G+\lambda)\,a_x\delta_{ix}\,. \qquad (4.3.13)$$

Substitution of Eqs. (4.3.4), (4.3.12) and (4.3.13) into Eq. (4.3.2) yields

$$(\rho_0 V^2 - G)\,a_i + \frac{1}{\mu}\,e_{ijk}\,e_{jlm}\,e_{mrs}\,n_l\,n_r\,B_k\,(e_{suv}B_u a_v + \xi a_x \delta_{sx})$$

$$-(G+\lambda)\,a_x\delta_{ix} = 0 \qquad (i = x, y, z)\,. \qquad (4.3.14)$$

This constitutes a set of three linear, homogeneous equations in a_x, a_y, a_z. Without loss of generality, we may assume $B_3 = 0$. The three equations read then, explicitly, with $n_x = 1$, $n_y = n_z = 0$,

$$(4.3.15)\quad \begin{cases} (\rho_0 V^2 - 2G - \lambda - \frac{1}{\mu} B_2^2) a_x + \frac{1}{\mu} B_1 B_2 a_y = 0 \\ \frac{1}{\mu} B_1 B_2 a_x + (\rho_0 V^2 - G - \frac{1}{\mu} B_1^2) a_y = 0 \\ (\rho_0 V^2 - G - \frac{1}{\mu} B_1^2) a_z = 0 . \end{cases}$$

The last equation represents a transverse acceleration wave, propagating with speed

$$(4.3.16)\quad V = \sqrt{\frac{1}{\rho_0}(G + \frac{1}{\mu} B_1^2)} = V_T \sqrt{1 + \frac{B_1^2}{\mu G}} .$$

The first two equations (4.3.15) for a longitudinal and a transverse wave are coupled. Putting their determinant equal to zero, one gets the following quadratic equation

$$(\rho_0 V^2)^2 - \left[3G + \lambda + \frac{1}{\mu}(B_1^2 + B_2^2)\right]\rho_0 V^2 + (2G + \lambda) G + \frac{G}{\mu}(2B_1^2 + B_2^2) + \frac{\lambda}{\mu} B_1^2 = 0 .$$

(4.3.17)

Both roots $\rho_0 V^2$ are real and positive. Since the coefficient ς drops out of the equation, one notes that the speed of propagation V , for the linearized plane wave approximation, is not influenced by the temperature field.

For a purely transverse magnetic field, $B_1 = 0$, we obtain a longitudinal wave with

$$V = \sqrt{\frac{1}{\rho_0}\left(2G + \lambda + \frac{B_2^2}{\mu}\right)} = V_L \sqrt{1 + \frac{B_2^2}{\rho_0 \mu V_L^2}} \qquad (4.3.18)$$

not coupled to the shear wave, cf. Section 4.2. For a purely longitudinal magnetic field, $B_2 = 0$, we have a shear wave with speed given by Eq. (4.3.16), not coupled to the longitudinal wave.

References

[1] R. Becker: Zur Theorie der Magnetisierungskurve. Z.f. Physik 62 (1930), 253.

[2] L. Knopoff: The interaction between elastic wave motions and a magnetic field in electrical conductors. J. Geophys. Res. 60 (1955), 441.

[3] R.A. Toupin: The elastic dielectric. J. Rational Mech. Anal. 5 (1956), 849.

[4] R.A. Toupin: A dynamical theory of elastic dielectrics. Int. J. Engng Sci. 1 (1963), 101.

[5] W.F. Brown, Jr.: Magnetoelastic Interactions. Springer-Verlag. Berlin-Heidelberg-New York 1966.

[6] H.F. Tiersten: Coupled magnetomechanical equations for magnetically saturated insulators. J. Mathematical Phys. 5 (1964), 1298.

[7] H.F. Tiersten: On the nonlinear equations of thermo-electroelasticity. Int. J. Engng Sci. 9 (1971), 587.

[8] A.C. Eringen: On the foundations of electroelastostatics. Int. J. Engng Sci. 1 (1963), 127.

[9] N.F. Jordan and A.C. Eringen: On the static nonlinear theory of electromagnetic thermoelastic solids. Part I and II. Int. J. Engng Sci. 2 (1964), 59 and 97.

[10] J.A. Stratton: Electromagnetic Theory. McGraw-Hill Book Comp., New York and London 1941.

[11] C. Truesdell and R.A. Toupin: The Classical Field Theories. In: Handbuch der Physik (Herausgegeben von S. Flügge). Bd.III/1. Springer-Verlag, Berlin 1960.

[12] C. Truesdell and W. Noll: The Non-Linear Field Theories of Mechanics. In: Handbuch der Physik (Herausgegeben von S. Flügge). Bd. III/3. Springer-Verlag, Berlin 1965.

[13] L.D. Landau und E.M. Lifschitz: Elektrodynamik der Kontinua. Akademie-Verlag, Berlin 1967.

[14] H. Parkus: Thermoelastic equations for ferromagnetic bodies. Arch. Mech. Stos. (to appear).

[15] A.E. Green and R.S. Rivlin: On Cauchy's equations of motion. Z. ang. Math. Phys. 15 (1964), 290.

[16] R.D. Mindlin: Polarisation gradient in elastic dielectrics. Int. J. Solids Structures 4 (1968), 637.

[17] H. Parkus: Thermoelasticity. Blaisdell Publishing Company. Waltham-Toronto-London 1968.

[18] S. Kaliski and W. Nowacki: Thermal excitations in coupled fields. In: Progress in Thermoelasticity (Editor W.K. Nowacki) Warschau, (1969).

[19] A.J. Willson: The propagation of magneto-thermo-elastic plane waves. Proc. Camb. Phil. Soc. 59 (1963), 483.

[20] G. Paria: Magneto Elasticity and Magneto-Thermo- Elasticity. In: Advances in Applied Mechanics 10 (1967), 73.

[21] M. F. McCarthy: Wave propagation in nonlinear magneto-thermo-elasticity. Propagation of acceleration waves. Proc. Vibr. Problems 8 (1967), 337.

Contents

Printed in the United States
By Bookmasters